Alan McKirdy has written many popular books and book chapters on geology and related topics and has helped to promote the study of environmental geology in Scotland. His other books with Birlinn include *Set in Stone: The Geology and Landscapes of Scotland* and *Land of Mountain and Flood*, which was nominated for the Saltire Research Book of the Year prize. Before his retirement, he was Head of Knowledge and Information Management at Scottish Natural Heritage. Alan is now a freelance writer and has given many talks on Scottish geology and landscapes at book festivals and other events across the country.

Mull, Iona and Ardnamurchan

LANDSCAPES IN STONE

Alan McKirdy

BIRLINN

For Moira

First published in Great Britain in 2017 by
Birlinn Ltd
West Newington House
10 Newington Road
Edinburgh
EH9 1QS

www.birlinn.co.uk

ISBN: 978 1 78027 440 9

British Library Cataloguing-in-Publication Data
A catalogue record for this book is available
on request from the British Library

Designed and typeset by Mark Blackadder

FRONTISPIECE.
Isle of Staffa.

Printed and bound in Britain by Latimer Trend, Plymouth

Contents

Introduction   7

Mull, Iona and Ardnamurchan through time   8

Geological map   10

1.   Time and motion   11

2.   Ancient foundations   14

3.   A hotter world   20

4.   Mull and Ardnamurchan volcanoes   22

5.   The Ice Age   33

6.   After the ice   36

7.   Landscapes today   39

8.   Places to visit   44

Acknowledgements and picture credits   48

Introduction

Lying off the south-western tip of Mull, the island of Iona has huge significance as the 'cathedral of the isles' and the first important centre of Christianity to be established in Scotland. But Iona Abbey itself is built upon rocks of much greater antiquity. The Lewisian gneisses of the western part of Iona (and also Coll and Tiree), formed between 2 and 3 billion years ago, are some of the oldest rocks to be found anywhere in the British Isles or indeed the world. These banded, and in places intensely folded, rocks were witness to some of the earliest events that shaped Planet Earth. They are known to geologists as 'basement' – the rocks on which all younger rock strata were subsequently overlain.

Opposite.
Calgary Bay, Isle of Mull.

As is usually the case, the rocks in this area do not give a complete record of the significant events that shaped the geological development of the whole landmass that was to become Scotland. A chasm in geological time has to be bridged before the next event around 800 million years ago is recognised from indelible evidence in the rocks of Mull and the surrounding area. A small patch of rock, known as the Moines, butts against pinkish-hued Ross of Mull granite, which was the next addition to the bedrock, around 400 million years ago. The Moines make up much of mainland Scotland to the north and west of the Great Glen, but on Mull this rock has just a cameo role.

After another significant lapse in geological time, sediments were laid down that tell of desert conditions during the Triassic period around 200 million years ago, and later, during the Jurassic, sediments were deposited under shallow seas. Subsequently, during the Palaeogene, two major volcanoes erupted that rocked the landscape.

The final event was the coming of the Ice Age. For over 2 million years, glaciers and ice sheets waxed and waned, as the climate oscillated from deep freeze to conditions similar to that of today. The passage of the ice ripped away much of the upper part of the two great volcanoes and gently scoured the landscape of the whole area to create the rounded contours that we see today.

Mull, Iona and Ardnamurchan through time

Period of geological time	Millions of years ago	Scotland's global position	Environments and events in Mull, Iona and Ardnamurchan
Anthropocene	Last 10,000	57° N	This is the time of *Homo sapiens* – humankind. Our ancestors have occupied these western outposts for five millennia now. During that time, the landscape has changed significantly to make life more tolerable for the growing number of inhabitants. In post-Industrial Revolution times, our effect on the planet – land, air and sea – has intensified, and that process of change will continue.
Quaternary	Started 2 million years ago	Present position of 57° N	• 11,500 ago to the present day – the ice finally melted and sea levels rose in response. River gravels and layers of peat accumulated in the benign conditions. • 12,500 to 11,500 years ago – the last local glaciation of the high ground on Mull occurred. When the ice melted, landslips formed throughout the area. • 14,700 years ago – this was a brief warmer interlude when temperatures were similar to those of today. The ice melted, and sea levels were higher than today. • 29,000 to 14,700 years ago – the area was entirely covered by thick ice. Glacial till accumulated under the ice. • Before 29,000 years ago – there were several advances of the ice, interspersed with warmer interludes.
Neogene	2–24	55° N	Climate cooled significantly and the Ice Age began. Prior to that the climate was subtropical.
Palaeogene	24–65	50° N	The North Atlantic Ocean continued to widen as North America and Europe moved further apart. During this period two major volcanoes were active in the area, erupting huge volumes of lava which created a new landscape.
Cretaceous	65–142	40° N	Warm shallow seas covered much of the land across Planet Earth with sea levels over 200m higher than today. This was a greenhouse world with higher levels of carbon dioxide in the atmosphere than current concentrations. Small patches of Cretaceous rocks have been found at Gribun and Carsaig on Mull.

Period of geological time	Millions of years ago	Scotland's global position	Environments and events in Mull, Iona and Ardnamurchan
Jurassic	142–205	35° N	Thick layers of sandstones, limestones and shales were laid down under a warm shallow sea that teemed with life, such as ammonites, corals and oysters.
Triassic	205–248	30° N	The area lay close to the Equator during these times. Desert conditions prevailed, with seasonal floods sweeping across the area which deposited thin layers of sands, silts and pebbly gravels.
Permian	248–290	20° N	Desert conditions were widespread across the land that would become Scotland, but no evidence is preserved on Mull, Iona, Ardnamurchan, Coll or Tiree.
Carboniferous	290–354	On the Equator	The area sat astride the Equator at this time. Tropical forests were widespread elsewhere across 'Scotland', but there is no geological record of these events here.
Devonian	354–417	10° S	The area formed a small part of a supercontinent that was located just south of the Equator. Conditions were hot and arid. A small patch of lava from these times is preserved north of Loch Spelve.
Silurian	417–443	15° S	The Ross of Mull granite was added to the bedrock. The Great Glen Fault was initiated and sliced through the southern edge of Mull.
Ordovician	443–495	20° S	The Iapetus Ocean began to close. This process eventually gave rise to the Highlands of Scotland.
Cambrian	495–545	30° S	The Iapetus Ocean widened.
Proterozoic	545–2,500	Close to South Pole	1,000 to 870 million years ago rocks were deposited that later gave rise to the Moine schists.
Archaean	Prior to 2,500	Unknown	• 2,800 to 1,700 million years ago – the 'basement' of Lewisian gneisses was created by alteration of even older rocks. • 4,540 million years ago – Planet Earth formed.

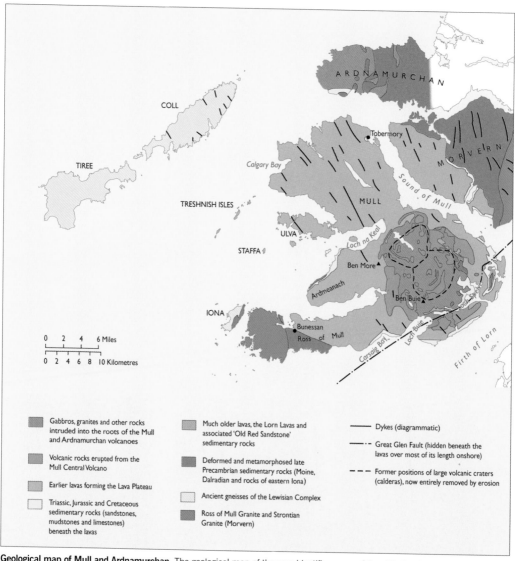

COLL

TIREE

ARDNAMURCHAN

Tobermory

Calgary Bay

M O R V E R N

Sound of Mull

MULL

TRESHNISH ISLES

ULVA

STAFFA

Loch na Keal

Ben More▲

Ardmeanach

Ben Buie▲

Loch Spelve

IONA

Bunessan

Ross of Mull

Loch Buie

Carsaig Bay

Firth of Lorn

0   2   4   6 Miles

0 2 4 6 8 10 Kilometres

▨	Gabbros, granites and other rocks intruded into the roots of the Mull and Ardnamurchan volcanoes
▨	Volcanic rocks erupted from the Mull Central Volcano
▨	Earlier lavas forming the Lava Plateau
▨	Triassic, Jurassic and Cretaceous sedimentary rocks (sandstones, mudstones and limestones) beneath the lavas

▨	Much older lavas, the Lorn Lavas and associated 'Old Red Sandstone' sedimentary rocks
▨	Deformed and metamorphosed late Precambrian sedimentary rocks (Moine, Dalradian and rocks of eastern Iona)
▨	Ancient gneisses of the Lewisian Complex
▨	Ross of Mull Granite and Strontian Granite (Morvern)

——— Dykes (diagrammatic)

–·–·– Great Glen Fault (hidden beneath the lavas over most of its length onshore)

– – – Former positions of large volcanic craters (calderas), now entirely removed by erosion

Geological map of Mull and Ardnamurchan. The geological map of the area identifies some of the oldest rocks in Scotland and also some of the most recent. Coll, Tiree and the western slice of Iona date back over 2 billion years. They formed part of an ancient continent of which Greenland and North America were close neighbours. Small areas of the next oldest rocks are located in the south-west of Mull, along with the Ross of Mull granite. These were the ancient layers upon which the younger rock strata were later stacked. As Mull and Ardnamurchan drifted northwards, the landmass lay for a few million years just to the north of the Equator. A thin skim of desert sands and muds, deposited under a shallow sea, accumulated on top of the ancient basement rocks. Next came the eruption of the Mull and Ardnamurchan volcanoes, which had an enduring impact on the land. Ancient magma chambers (now solidified) and associated erupted lava fields were moulded and smoothed by the much later passage of ice into the familiar landscapes we see today.

1
Time and motion

Time

The caves along the south side of Loch na Keal on Mull were once home to some of the earliest settlers to have inhabited the island. Between 7,000 and 5,000 years ago, nomadic hunters and fishermen made Mull their home. In later Neolithic times, settlers created elaborate burial mounds from rock and earth as a link between this world and the next. Standing stones, stone circles and cup-and-ring markings are enigmatic and mysterious evidence of this prehistoric settlement.

Our human history seems to represent the edge of time, but we can extend that timescale backwards millions of years through the study of rocks and landscapes. The rocks of Coll, Tiree and Iona contain evidence of some of the earliest events in Scotland's geological

Standing stones at Dervaig, Mull.

11

history: some of these rocks are over 2 billion years old. Lengths of time such as this are difficult to make sense of, because we tend to understand things in terms of human timescales. But an essential part of the study of geology is learning to appreciate this extended timescale.

Accurate methods of dating rocks were developed during the 1950s and 1960s by Professor Arthur Holmes, who worked at Edinburgh University. His method relied on an understanding of the way in which naturally occurring radioactive minerals decayed and the timescale over which that process took place. For the very first time, this allowed an absolute date to be placed on rock formations. The ages of rocks from different parts of the world could then be compared. Our appreciation of the events that shaped the rocks and landscapes of Mull, Iona and Ardnamurchan are considerably enhanced by this better understanding of geochronology – the accurate dating of rocks and geological events.

Opposite.
Sanna Bay, Ardnamurchan.

Motion

Change has been a constant feature throughout geological history. Planet Earth is a dynamic place where the outer layer, known as the crust, comprises a series of individual plates that move and jostle their way across the globe. So the distribution of land and sea is constantly changing. Energy emanates from the Earth's core and creates movement in the overlying mantle that carries the plates along at a rate of between 3cm and 6cm per year. Given sufficient time, these forces can move the continents to the other side of the globe and collide with other landmasses on their journey to create new mountain ranges. This understanding of how the Earth works is known as plate tectonics.

Heat flows from the Earth's core, which is a raging 6,000°C. This sets up a convection cell that moves the tectonic plates that comprise the Earth's crust. The crust is the outer layer shown in green.

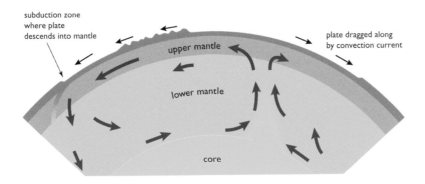

subduction zone where plate descends into mantle

upper mantle

plate dragged along by convection current

lower mantle

core

2
Ancient foundations

The islands of Coll and Tiree, along with the western slice of Iona, have the most ancient rocks of any of the islands in this area. This bedrock, composed of Lewisian gneiss (pronounced 'nice'), is a fragment of some of the very oldest rocks preserved on Planet Earth. Other slabs of Lewisian gneiss are found in the Western Isles and areas of the mainland, notably around Loch Maree, Lochinver and Cape Wrath. Together they represent the very earliest time in the development of the land that was to become Scotland. These rocks date back beyond 2,000,000,000 years, so some of the finer detail of their origin and formation has inevitably been lost. But we do know that these rocks have had a long history of periodic burial and heating at incredible depths in the Earth's crust, which has fundamentally changed their character. They have been deformed in this hostile environment by squeezing, partial melting, buckling, folding and faulting in a process known as metamorphism. Subsequently a combination of earth movements and erosion by ice, wind and water skimmed the overlying strata away to reveal the banded, gnarled rocks we see at the surface today.

The ancient rocks of Coll and Tiree were first mapped by the Geological Survey of Great Britain in 1921, with the results published in a detailed account in 1930. One of the surveyors was Sir Edward Battersby Bailey, a stalwart of the Geological Survey. He was twice wounded during the first World War, losing his left eye and the use of his left arm; he received the Military Cross. The men, and they were exclusively men, who undertook these commissions to map the remoter parts of the country, were clearly made of stern stuff.

Bailey and his colleagues found the Lewisian gneisses to be composed mainly of greatly altered metamorphic rocks, often partially melted by intense heat, with small patches of more exotic rock strata that were originally sediments. Both on Coll and Tiree, thin bands of marble, some decorative in nature, were identified by these early pioneers. Tiree marble from Balephetrish Bay is perhaps the most famous example of this unusual rock. Rocks of similar antiquity are

Sir Edward Bailey was a Cambridge graduate who rose to become Director of the British Geological Survey. He was Professor of Geology at Glasgow University prior to this appointment. One of his many assignments with the Geological Survey was to map the bedrock of Coll.

also found on Iona, where Lewisian gneisses make up the rocky terrain on the western part of the island.

The gnarled and contorted nature of the Lewisian gneisses is beautifully demonstrated here on Iona.

Marble is also to be found on the south coast of Iona. These marbles were originally laid down as sediments on an ancient lake or sea bed. As a result of repeated episodes of heating and squashing deep within the Earth's crust, they are now streaked and banded in appearance.

The Ross of Mull granite on the western extremity of the island, is also part of the ancient suite of rocks that forms the foundations of Mull. It dates back to around 420 million years and was formed as part of the intense activity related to the creation of the Caledonian Mountains. The area now recognised as Mull was on the periphery of these tumultuous events, but clearly affected by them.

This pink coarse-grained granite is ideal for construction. It has widely spaced cracks or joints, which allowed substantial blocks to be quarried and used in a wide variety of iconic structures. The lighthouses at Ardnamurchan and Skerryvore were constructed from this material as were parts of Iona Abbey. Further afield, Holburn Viaduct and Blackfriars Bridge in London, and the docks at Liverpool,

Iona Abbey sits close to the Lewisian gneisses on the western part of the island. The flatter land between the abbey and the sea is underlain by strata of a slightly younger vintage that had previously been identified as Torridonian sandstones. But their origin is now considered to be less certain as they have been much altered since they were first deposited as sediments.

Glasgow and New York, were all constructed, either wholly or in part, from Ross of Mull granite.

Another short volcanic burst occurred around the same time as the Ross of Mull granite was added to the bedrock of the island. The land was still unstable from the tumultuous events that had led to the formation of this new arrangement of continents. Earthquakes and volcanic activity were frequent. A small patch of basalt lava is preserved just to the north of Loch Spelve as evidence of this instability. At this time, proto-Scotland formed part of a continent that lay 10° south of the Equator. The area was landlocked, arid and bare of vegetation. Plants of significant size had yet to evolve.

This abandoned quarry on the south coast of Iona was worked for its marble.

The Ross of Mull granite gives rise to a very distinctive landscape of rolling knolls interspersed with heather cover and croftland. In the foreground is land underlain by granite and in the distance are the lavas of the Ardmeanach Peninsula with the head of Loch Scridain beyond.

Right.
A close-up view of the
Ross of Mull granite.

Below left and right.
These two iconic buildings
are built using Ross of Mull
granite – the Ardnamurchan
lighthouse and Blackfriars
Bridge in London.

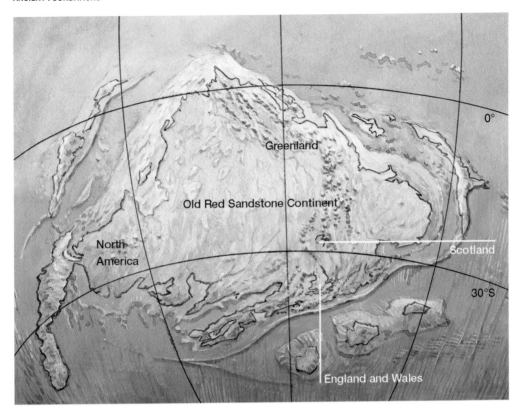

Greenland

Old Red Sandstone Continent

North
America

Scotland

0°

30°S

England and Wales

Mind the gap

A period of around 100 million years elapsed before the next geological event left a permanent mark on the landscape. These periods of hiatus, where substantial gaps in the record of the rocks occur, are not unusual in Scotland or indeed around the world. Significant geological events were happening to the east in what became mainland Scotland. Most noteworthy of these events was the formation of tropical rainforests between what are now Scotland's two principal cities – Edinburgh and Glasgow. During Carboniferous times, 'Scotland' lay astride the Equator, and an exotic rainforest ecosystem clothed the landscape. But there is no hint of these events in the record of the rocks on Mull, Iona, Ardnamurchan, Coll or Tiree. Neither are the deserts sands that covered much of central Scotland during Permian times represented here. There may once have been evidence for these events and environments in this area but, if it did exist, it is now long gone, wiped away by subsequent erosion.

Around 400 million years ago, the land that was to become Scotland was a parched land, but flash floods and lakes of significant size are known to have existed. It was part of a new landmass, known as the 'Old Red Sandstone Continent', which was a crazy melange of different fragments of crust that are now separated by large distances and, in some cases, wide oceans. North America sat alongside Greenland and the land that was to become the British Isles. It was literally a different world 400 million years ago.

3
A hotter world

We pick up the story again in Triassic times, when Scotland had drifted significantly further north from the Equator. Although still hot and arid, rivers flowed across the parched land, dumping thick layers of rounded boulders in a deposit that we now describe as a conglomerate. Other rocks dating from this time are thought to have accumulated in a shallow lagoon that periodically dried out in the tropical heat. These rocks from Triassic times were dumped directly on top of the ancient basement rocks, and the line that separates the two represents around 700 million years of geological time.

The rock strata may, at one time, have contained evidence of the plants and animals that inhabited this hostile world, but no significant fossil remains have been recovered to date, despite the area having been first surveyed in 1925. The presence of very thin fossil soils has also been noted. The best place to see these deposits is at Gribun shore, close to Inch Kenneth, Mull.

The cliffs at Gribun on Mull show an upward succession from the very oldest basement rocks through younger strata (Triassic and Cretaceous) which are capped by a thick sequence of lavas that flowed from the Mull volcano.

Palaeogene lavas

Upper Cretaceous

Upper Triassic

Moine

Moine

Moine

At the end of the Triassic Period, sea levels rose worldwide and clear evidence for that global event is preserved on Mull. Lying above the Triassic sediments that were deposited on land by rivers and lakes are strata laid down under the sea. Careful study has allowed geologists to say that the oldest Jurassic deposits, known as the Lias, originated in waters that varied from shallow near-shore conditions to slightly deeper waters. Ammonites are occasionally uncovered in these layers. In later Jurassic times, thick deposits of clay were laid down, probably in deeper waters, indicating that sea levels were continuing to rise at this time.

Following the end of the Jurassic Period, sea levels rose to record levels during Cretaceous times. Evidence suggests that polar ice caps completely melted and the planet was entirely ice-free. Water locked up in the ice caps for much of geological history was liberated, causing this substantial rise in sea level. Today, we talk in panic-stricken tones about rises in sea level of around 0.5m over the next 100 years. But during Cretaceous times, sea levels were around 200–300m higher than they are today. This was a greenhouse world with high carbon dioxide levels and shallow seas covering most of the world's continents.

Only small patches of Cretaceous rocks have been found on Mull, notably at Gribun and Carsaig. Strata of Cretaceous age are emblematic of southern England, where the chalk of the white cliffs of Dover is celebrated in paintings and song. Although not prompting such an outbreak of patriotic fervour, thin layers of chalk up to 3m thick have been noted at Gribun.

Above.
Sedimentary layers of Jurassic age make up the lower parts of the cliff on the far side of Carsaig Bay, Mull.

Below.
A fossil ammonite of Jurassic age recovered from the Jurassic strata at Carsaig Bay, Mull.

4

Mull and Ardnamurchan volcanoes

Again, we must look to global events to understand the next sequence of events that left an imprint on Mull and Ardnamurchan. And it was a sizable imprint. The continent of which Scotland was a small part had started to break up, with energy from deep within the planet focused like a blowtorch on the underside of the Earth's crust.

Ferocious blasts of super-heated rock, known as mantle plumes, welled up from deep below the surface. These events marked the opening of the Atlantic Ocean and caused the demise of Pangaea.

This process released huge quantities of heat. The main manifestation of this transfer of energy from the bowels of the Earth to the surface was the creation of many active volcanoes. St Kilda, Skye, Rum, Arran, Ailsa Craig and, of course, Mull and Ardnamurchan were all active as a result of this process of continental break-up.

The Mull volcano was a hugely imposing structure. What we see today is just the eroded roots of what was formerly a much bigger edifice. But what remains has allowed geologists to piece together a

With the passage of many millions of years, continents continued to move across the globe, powered by the currents in the Earth's mantle. During Triassic times, around 250 million years ago, these movements created a super-continent that stretched from pole to pole. This is known as Pangaea – 'All Earth'. But later, cracks appeared in Pangaea as the continent started to break up and its component parts drifted apart to be later separated by the North Atlantic Ocean.

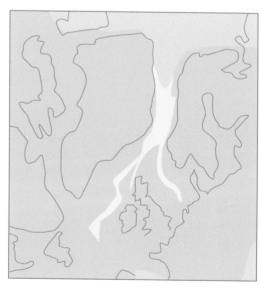

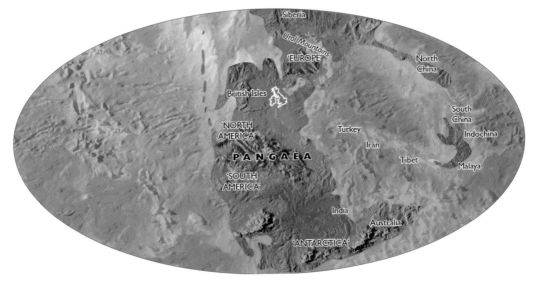

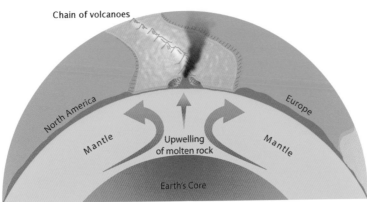

Above.
Pangaea, which contained Scotland, spanned the globe from pole to pole. But continental alliances don't last forever and this odd mix of crustal fragments was soon to be dispersed as the North Atlantic Ocean started to open.

Left.
The North American and European continents were forced apart as the mantle plume created a new ocean floored by basalt lavas. New molten material has been added along the spine of the ocean from that time to this, forcing the continents to move further apart. This process initiated and energised a line of volcanoes running down the north-western coast of Scotland from St Kilda to Arran, including Mull and Ardnamurchan.

picture of what it would have looked like in its prime and how its internal plumbing would have functioned.

The first burst of volcanic activity was the eruption of copious quantities of basalt lava starting around 60.5 million years ago. These outpourings were to last for almost 2 million years without any discernible break. The dates for these are very similar to those of the Skye volcano to the north, so both would have been active around much the same time.

The lava flows gushed from a crack (or fissure) in the ground that ran nearby. There is little associated volcanic ash, indicating that the lava just oozed from the ground without much explosive activity. Individual lava flows can be around 15 metres thick, most containing

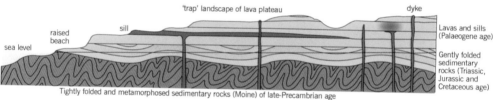

'trap' landscape of lava plateau — dyke

sill — raised beach — Lavas and sills (Palaeogene age)

sea level

Gently folded sedimentary rocks (Triassic, Jurassic and Cretaceous age)

Tightly folded and metamorphosed sedimentary rocks (Moine) of late-Precambrian age

Top.
The pile of lavas forming the Ardmeanach peninsula built slowly, layer upon layer, with each eruption. The lavas were runny and spread out over the top of the previously erupted layers.

Above.
This cross-section through the lavas of Mull shows the lavas that were erupted across a landscape of older strata. These successive eruptions formed what is known as a 'trap topography' of steps and stairs which are a prominent feature of the landscape today.

cavities and gas bubbles, some filled by minerals such as agates and other attractive crystals.

The reconstruction of the lava field associated with the Mull volcano, on the basis of work undertaken by Sir E.B. Bailey and more recent studies, suggests that it extended to Morvern and Loch Aline on the mainland and northwards to form Muck and Eigg. So the dramatic skyline of the Sgurr of Eigg was carved from the lava flows that emanated from the distant island of Mull.

The Isle of Staffa, with its world-renowned columnar-jointed structure, also owes its origins to the Mull volcano.

Another iconic feature of the Mull lavas is MacCulloch's fossil tree on the Ardmeanach peninsula. Over 12 metres high and around 60 million years old, this impression of a tree embedded in old lava flow was discovered by John MacCulloch in 1819.

Further evidence of the plant life that thrived between successive

Left.

This map shows the extent of the Mull and Ardnamurchan volcanoes and their component parts – the magma chambers and the associated lava fields. The magma chambers are shown in green, the onshore lava fields in pink and the offshore extent of the lavas in purple.

Below.

This dramatic view of the Sgurr of Eigg symbolises this island. The rock type is unusual – a pitchstone some 120 metres thick. This lava is acid in composition and would have flowed slowly across the ancient landscape as it was viscous, like treacle. The lava sits on a layer of conglomerate laid down by a fast-flowing stream. There was sufficient time for woodlands to become established between eruptions of the volcano and here we see evidence that rivers of substantial size flowed across the landscape, depositing thick layers of boulders, sands and muds in layers characteristic of water-lain sediments.

The distinctive columns of lava flows that created Staffa were formed as the molten rock cooled and shrank into these characteristic six-sided columns. Staffa has inspired music and art: Felix Mendelssohn wrote the *Hebrides Overture* after his visit in 1829 and J.M.W. Turner, who visited a year later, exhibited a painting of Staffa in 1832.

lava flows is provided at Ardtun near the western extremity of Mull. Thin soils of reddish hue developed between eruptions and these layers have yielded a variety of plant remains, including leaves from oak, hazel, plane, magnolia and ginkgo. Other tree species have been identified by the signature fossil pollen grains that have also been recovered from these rock strata.

Elsewhere, some thin coals have been identified. This indicates that low-lying swampy areas developed on top of the last eruption of lava flow, and this wet environment supported a thriving woodland ecosystem. The variety of tree species that grew in this swamp suggests a warm temperate climate similar to that of the Mediterranean today.

As the eruption of lavas diminished, the magma chamber under the volcano had been substantially emptied and a circular crack developed at the surface. The lavas inside the ring of the fault then subsided into the void below under their own weight. There are indications that a crater lake developed in the central sunken area of the volcano. The volcano continued to grow as new molten rock was pumped into the magma chamber from below.

The circular ring fault allowed molten rock to pulse closer to the surface. One dramatic episode created a ring dyke (a circular intrusion of volcanic rock) at Loch Ba towards the north-west of the magma chamber. Granite magma was squirted into place when a chunk of

Left.
MacCulloch's tree. Although little remains of this large conifer that was engulfed by lava flows, it demonstrates that an extensive woodland ecosystem sprang up between the eruptions of successive lava flows. But with each new eruption of molten rock, any signs of life were extinguished as the lavas scorched their way across the landscape.

Below.
A fossil hazel-like leaf has been recovered from a fossil soil at Ardtun.

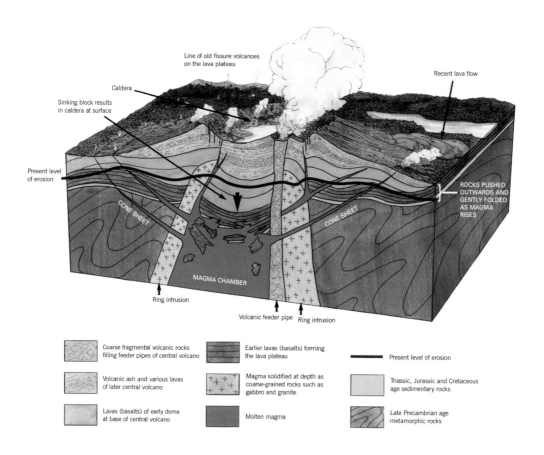

Line of old fissure volcanoes
on the lava plateau

Recent lava flow

Caldera

Sinking block results
in caldera at surface

Present level
of erosion

ROCKS PUSHED
OUTWARDS AND
GENTLY FOLDED
AS MAGMA
RISES

CONE SHEET

CONE SHEET

MAGMA CHAMBER

Ring intrusion

Volcanic feeder pipe   Ring intrusion

Coarse fragmental volcanic rocks filling feeder pipes of central volcano	Earlier lavas (basalts) forming the lava plateau	Present level of erosion
Volcanic ash and various lavas of later central volcano	Magma solidified at depth as coarse-grained rocks such as gabbro and granite	Triassic, Jurassic and Cretaceous age sedimentary rocks
Lavas (basalts) of early dome at base of central volcano	Molten magma	Late Precambrian age metamorphic rocks

This 3D reconstruction of the Mull volcano is a compilation of the available evidence collected from the hills and glens of Mull over the last century. It shows the internal workings of what would have been a long-lived and very active volcanic centre. The caldera collapsed into the magma chamber below, causing a small depression at the surface. It also shows thin sheets of molten rock that had been squirted laterally from the magma chamber to form structures known as cone sheets.

crust collapsed downwards into the magma chamber, displacing molten rock upwards, exploiting the existing weakness of the ring fault. This rather unusual type of eruption created a ring of granite with a diameter of around 8 kilometres and a maximum thickness of 400m. A cross-section through this ring dyke intrusion is shown in the diagram above.

Periodic larger eruptions then took place, and the central portion of the volcano further collapsed along the circular trace of the fault line. This is called a caldera eruption. Present-day caldera eruptions are noted for their explosive nature, and when the Mull volcano blew its top it would also have done so with extreme violence.

Volcanic eruptions continued, and further lavas erupted from the magma chamber. These lavas had a limited geographic spread and were concentrated mainly around the central core of the volcano.

The Ardnamurchan volcano

Today, no lavas survive that were associated with the eruption of the nearby and entirely separate Ardnamurchan volcano. It could be that there were none in the first place or that they were entirely removed by subsequent erosion by ice, wind and water. This volcano first erupted around 60 million years ago and remained active for about 2 million years. After the ice cut deep and removed about 2 kilometres of its superstructure, we were left with the magma chamber. What is particularly remarkable about the Ardnamurchan volcano is the symmetry of its roots.

This place is one of the wonders of Scotland. It is geology writ large in the landscape with little greenery and few buildings to obscure the essential natural elements. Pressure from the magma chamber below created a series of circular faults that became a pathway for molten rock. The analogy of a full wine bottle in which the cork is pressed down and liquid flows upwards and around the cork has been used to help explain this phenomenon. Successive pulses of molten rock then create the circle next to the previous circle. The main focus of the volcanic activity changed over time, with three centres recognised. The first cut across the second, and the third volcanic centre, known as Centre 3, is the last and final version that we see today.

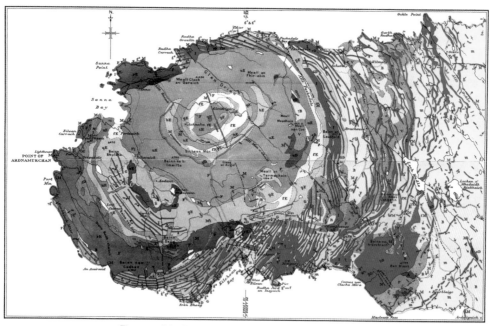

The map of the Ardnamurchan peninsula, prepared by the Geological Survey, which accompanied their description of the area, is a work of art. Published in 1930, it matches beautifully with the aerial view. The final pulse of magma that gave rise to what geologists call 'Centre 3' is the most significant and gives the area its distinctive amphitheatre feel.

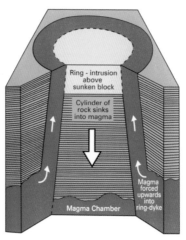

Formation of ring dykes and ring-intrusions above a magma chamber

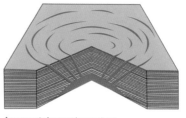

Arrangement of concentric cone sheets above a magma chamber

This is how the ring dykes of Ardnamurchan formed. The circular faults that developed were later exploited by molten rock. Cone sheets were injected in response to the pressure field that created wider-spaced circular cracks through the overlying rocks. They too were subsequently filled by molten rock from the magma chamber below.

Vertical sheets of magma

Throughout the life of these two volcanic centres, vertical sheets of molten rock, called dykes, were being forced into the surrounding bedrock and beyond. Many of these dykes are around 2 metres thick, but some are wider. What is most extraordinary about these structures is how far the molten rock travelled after being forced out of the volcano. The Mull volcano propelled thin linear sheets of molten rock in south-easterly and easterly directions across the Scottish mainland.

The furthest travelled is a dyke found in North Yorkshire which is a staggering 400 kilometres from Mull! It is thought to have been a single pulse of molten rock that exploited pre-existing weaknesses in the rocks through which it travelled. Its course from Mull to Whitby was more or less a straight one, but the force with which this material was propelled was so large that it is difficult to comprehend. It had to remain molten throughout its 400km journey. And we know how quickly this linear streak of molten rock would have cooled, so it has been calculated that the dyke would have forced its way through the host rock at around walking pace.

This linear ribbon of rock was forced out under high pressure into the surrounding rocks in a molten state, then cooled quickly. It stands proud today as it is harder than the surrounding rocks.

Right.
This diagrammatic representation of the dykes that emanated from the Mull volcano shows their extent across Argyll and adjacent areas and the distance these pulses of molten rocks travelled from the central volcano.

Below.
This dyke was first described by the prominent geologist James Hutton on his travels across Scotland in the mid eighteenth century. It is around 130 kilometres east of Mull, but it is thought to be an offshoot of the Mull volcano.

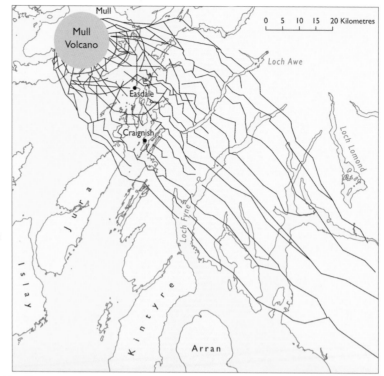

5
The Ice Age

After the temperate climate of the Palaeogene Period when the volcanoes were active, temperatures declined gradually as the Ice Age took hold. Over a period of around 2.6 million years (up to the present day), the climate fluctuated between periods when temperatures were

This image of Antarctica shows how Scotland might have looked during the last Ice Age. As the ice advanced and retreated, it had a significant impact on the landscape, shaping the mountains and glens we see today.

As the glacial ice melted, the debris that was picked up by the ice was dumped into jumbled piles. So the bedrock, which had been smoothed by the passage of the ice, was later plastered by a chaotic melange of boulders, sand and mud called glacial till. The landforms created in this way are called moraines, but the deposit is comprised of glacial till. This cover is many metres thick in places and entirely absent in others.

similar to those of today and times when thick layers of ice and snow covered the land throughout the year. Arctic conditions prevailed over much of continental Europe. As large volumes of water in seas and rivers became locked in glacial ice and snow, sea levels fell dramatically to new lows. The British Isles were, at this time, linked to continental Europe. In fact, the area now occupied by the North Sea was a large expanse of dry land supporting a human population estimated to have been around 100,000.

What caused these dramatic temperature fluctuations? The answer lies in our planet's relationship to the Sun. The Earth's orbit around the Sun is not constant. Over a period of about 100,000 years it changes gradually from a circular orbit to a more elliptical one, and then repeats this cycle. When the planet is in an orbit that takes it furthest from the Sun, the Earth's surface receives less sunlight and this causes temperatures to drop. The angle of the Earth's tilt also varies, and this affects the amount of solar radiation that falls on the surface of the planet. The combined effect of these produces a sustained period of low temperatures, causing the climate to descend into a cold phase: an ice age. The fluctuations from cold phases to warmer phases have happened many times over the last 2.6 million

years and they will continue in the future. Currently we are in a short spell of warmer conditions called an inter-glacial period, but the glaciers will return before long, perhaps even in the next 50,000 years, and then we will be plunged again into another ice age.

About 22,000 years ago the last glaciation reached its peak when vast sheets of ice flowed from the highest ground towards the sea. Then temperatures rose suddenly around 14,700 years ago, melting the ice caps and, for about 1,800 years, summer temperatures were similar to those of today. Boulders, stones and mud carried by the ice were dumped as the ice melted and formed characteristic piles of glacial debris, known as moraines.

Pioneer plants such as grasses and heathland vegetation established themselves across the post-glacial landscapes in the warmer temperatures. But these more hospitable conditions were short-lived. Around 12,000 years ago the climate began to cool again and ice once again accumulated on higher ground. It was during this last glacial period that many of the landscape features that are characteristic of the area were developed.

This splendid perched boulder on Ben Hough, Coll, demonstrates the power of the ice to move rocks around. This is known as a glacial erratic – a lump of rock moved from one location and then unceremoniously dumped as the ice melted. This rock has been perched here for at least the last 12,000 years.

6
After the ice

A new beginning

The climate started to warm abruptly again around 11,500 years ago and the ice vanished quite quickly thereafter. A modern woodland ecosystem soon became established, and this was the world that people first encountered when they settled the area around 7,000 years ago. These early settlers modified their local environment by chopping down areas of woodland to grow crops, establishing an increasingly pastoral way of life.

Changes at the coast

The overall shape of the coastline has changed little since the ice melted, but sea levels have fluctuated considerably. Two factors were at work here. When temperatures rose and ice sheets melted, enormous quantities of water were added to the world's oceans. But the land itself also moved. For thousands of years during the Ice Age, the landmass was weighed down by thick layers of ice, so when the ice sheets melted, the land 'bounced' back, very slowly, to a higher level (sometimes tens of metres higher): a process known as isostatic readjustment.

Coastal features such as benches and notches high above the present high-water mark were cut into the cliffs by wave action. They clearly indicate that the sea level was once higher than it is today.

Unstable cliffs

The eruption of great piles of lava has long-term consequences for the stability of the land, particularly near the coast. The underlying rock strata on the coast of Mull are relatively weak in engineering terms, so the burden of the lava pile was difficult to support. The result is a series of landslips, which are particularly clear to see at Gribun. A jumbled

The work of the sea is clear on the cliffs at Treshnish Point on Mull. A wide bench has been cut in the bedrock by wave action, and a line of cliffs, once battered by Atlantic storms, shows a much earlier high water mark. Since that time, the sea level has fallen relative to the land.

Lava flows cap the underlying Cretaceous and older rock strata near Gribun on Mull, and the instability of the cliffs is clearly evident.

mixture of blocks of slumped rock have cascaded down the slope, and this instability is likely to continue for years to come.

Taking the long view of climate change

There is a strong consensus among the scientific community that climate change today is being driven by the actions of humanity, including the emission of greenhouse gases. But we should also take the long view of climate change, recognising that this is a dynamic planet and there is unassailable evidence for climate change in the geological past. This evidence is firmly locked in the record of the rocks. In addressing the issue of anthropogenic (or man-induced) climate change, we can learn a great deal from processes that have changed the climate in the past.

The rocks and landscapes of this area and many other parts of Scotland, and indeed the world, carry the imprint of past climatic regimes that are very different from those of today. If natural climate change is demonstrably a key part of our geological story, then there is every chance that variations in climate will be observed in the future. Today, the argument is largely about what effect our actions are having, and have had in the recent past, on the global climate. But in assessing the evidence for man-induced changes to our world, we should not forget the fact that a changing climate is an integral part of the way our dynamic planet functions.

7
Landscapes today

Machair landscapes

Machair is one of the rarest habitats in Europe, found only in the north and west of Britain and Ireland. It is a thin strip of fertile land between the inland rocks and the sea. The soil is made from shell fragments, broken up by the sea offshore and in the inter-tidal area, and then blown by the wind over the beach and inland. Areas of machair are extensive around the islands of Coll and Tiree, the western extremity of Ardnamurchan and on Iona.

A wide expanse of machair lies between two rocky knolls of Lewisian gneiss.

Aesthetic landscapes

Loch na Keal is formally designated as one of the National Scenic Areas of Scotland. An extract from the official description of this special place reads 'The shoreline of the inner loch is of low relief, backed by meadow and woodland, above which the slopes sweep uniformly up to the shapely peak of Ben More.'

Built landscapes

The waterfront at Tobermory is one of the most recognisable streets in Scotland, immortalised in the children's television programme *Balamory* (aka Tobermory!). The brightly painted shops and houses are iconic of Mull and a magnet for tourists. The museum, located on the sea front, has exhibits and information about the local geology and is well worth a visit.

Natural landscapes

On the island of Coll, conservation areas have been designated to protect internationally important populations of over-wintering Greenland white-fronted geese, which graze on the unimproved grasslands. Alongside these migrants are moorland and wetland breeding birds including merlin and red-throated divers. The open moorland is one of the most southerly stations supporting breeding populations

Opposite.
The shore and the land at Gribun on Loch na Keal form part of a National Scenic Area. The rock strata near the shoreline are of Jurassic age and older, capped by lavas from the Mull volcano. Ben More in the distance is built from the dissected remains of the volcano's magma chamber.

Below.
The Tobermory waterfront.

Above left.
Dunlin.

Above right.
Merlin.

Right.
Grey seal.

of Arctic skua. Also breeding here are hen harriers, dunlin, great skua and twite. Sea eagles were re-introduced to this ecosystem, and significant populations of these birds are now thriving in the area.

Ben More, Mull's highest peak, is designated as a Site of Special Scientific Interest because of upland oak woodland habitat. This woodland consists of oak, birch, hazel, rowan, holly and willow trees. The habitat supports a rich assemblage of mosses, liverworts and lichens for which the area is renowned.

The Treshnish Isles, lying just offshore from Mull, are important for their seabird populations and colony of grey seals, the largest in the UK. The low-lying rock platform and beach at the north end of Lunga and on Sgeir a'Chaisteil are ideal breeding grounds for this sea mammal. The cliff slopes and cliff-top habitats support a rich maritime grassland plant community including thrift, sea campion, common dog violet, bird's-foot trefoil and wild thyme.

Left.
Common dog violet.

Right.
Bird's-foot trefoil.

8
Places to visit

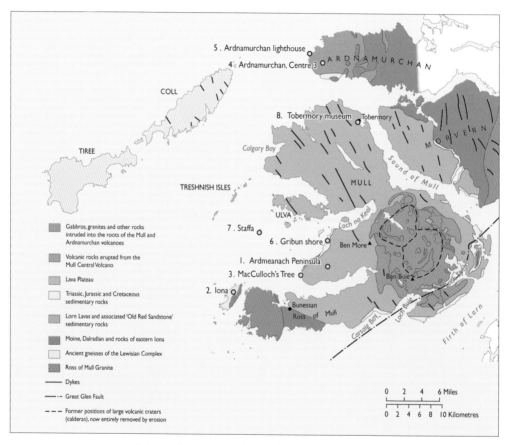

Many places to visit have already been described earlier in the text, so this short chapter just highlights some the best places to see the geology and landscapes of the area at first hand. The OS Landranger sheets that cover the area are 46, 47, 48 and 49.

Further details of many places of natural heritage interest (geological, biological and landscape) can be found on the SNHi portal at www.snh.gov.uk/snhi.

1. **Ardmeanach peninsula:** the stepped profile of the lavas associated with the Mull volcano are well demonstrated along the southern shore of the Ardmeanach peninsula. The road to Iona, the A849, which runs along the southern shore of Loch Scridain, provides some of the best vantage points.

2. **Iona:** the cultural and religious attractions of Iona are well known. The Lewisian gneisses can be examined on the west side of the island and the marble quarry at the south end. The ruins of the nunnery, just to the south of the abbey, demonstrate the geology of the island well, as a mix of granite, gneiss and sandstone has been used in its construction.

Top.
View across the Ardmeanach peninsula.

Above left.
View from Iona of the Ross of Mull and Ben More beyond.

Above right.
A wall of the nunnery at Iona, showing different building stones.

These images attempt to convey the amphitheatre form that lies at the heart of the Ardnamurchan volcano. The outer ring of rocks defining Centre 3 is well displayed along the skyline.

3. MacCulloch's tree: for the more adventurous, taking the B8035 along the northern shore of Loch Scridain and then striking off westwards along an unclassified track leads the visitor to the site of the fossil tree.

4. Ardnamurchan, Centre 3: the beautifully drawn map of the Ardnamurchan volcano, shown on page 30, depicts the complexity of the area. Travel along the B8007 road to this naturally excavated circular structure with its nearly complete ring of hills dominating the peninsula. Stand in the centre of the structure, turn 360° and be amazed by one of the wonders of Scotland. A visit here should be on everyone's geological bucket list.

5. Ardnamurchan lighthouse: moving away from the centre of the volcano to the coast, the area around the lighthouse exposes rocks of the outer circle of molten rock injected into the magma chamber.

6. Gribun shore: the description on page 20 gives a clear account of the variety of rocks exposed here, from the very oldest Moine rocks, through Triassic to the lavas of the Mull volcano. It's an interesting succession of strata. Access is from the B8035 road that runs along the southern shore of Loch na Keal.

7. Staffa: this island has many artistic and musical associations and is well worth a visit. Boat trips run from Oban and Iona.

8. Tobermory museum: this small museum on the main street of Tobermory has geological exhibits, making it worth a visit.

Left.
The interior of Fingal's Cave on Staffa, showing the columns of basalt.

Right.
Ardnamurchan lighthouse.

Acknowledgements and picture credits

Thanks are due to Professor Stuart Monro OBE FRSE and Moira McKirdy MBE for their comments and suggestions on the various drafts of this book. I also thank Andrew Simmons, Debs Warner, Mairi Sutherland and Hugh Andrew from Birlinn for their support and direction. Mark Blackadder's book design is up to his usual high standard. Scottish Natural Heritage, in association with the British Geological Survey, published the *Landscape Fashioned by Geology* series that was the precursor to the new *Landscapes in Stone* titles. I thank them both for their permission to use some of the original artwork and photography in this book. David Stephenson wrote the original text for *Mull and Iona – A Landscape Fashioned by Geology*, which influenced aspects of this book. I dedicate this book to my wife Moira to celebrate over forty years of marriage. She has tolerated my frequent visits to PreCambrian times largely with good humour. Her scientific knowledge and proofreading skills have proved to be invaluable in producing the *Landscapes in Stone* series.

Picture credits

Title page ARV/Alamy Stock Photo; 6 Spumador; 10 drawn by Jim Lewis; 11 Gail Johnson; 12 ScotImage/Alamy Stock Photo; 13 drawn by Jim Lewis; 14 Reproduced courtesy of the British Geological Survey EA17/056; 15 Sue Burton Photography Ltd; 16 Permit Number CP17/025 British Geological Survey © NERC 2017. All rights reserved. Also 21 (lower), 27 (right); 17 (upper and lower) Lorne Gill/SNH; 18 (upper) Lorne Gill/SNH (left) Lukassek (right) Kurt Pacaud; 19 drawn by Jim Lewis; 20 Brian Upton; 21 (upper) Michael MacGregor; 22 drawn by Jim Lewis; 23 (upper and lower) drawn by Jim Lewis; 24 (upper) Lorne Gill/SNH (lower) Craig Ellery; 25 (upper) drawn by Jim Lewis (lower) Alan Keith Beastall/Alamy Stock Photo; 26 Lorne Gill/SNH; 27 Colin MacFadyen ; 28 Clare Hewitt; 29 Patricia & Angus Macdonald/Aerographica/SNH; 30 (upper) Map of Tertiary Igneous Complex of Ardnamurchan, source *The Geology of Ardnamurchan, north-west Mull and Coll* by J.E. Bailey, H.H. Thomas, E.B. Bailey, B.E. Dixon, V.A. Eyles, G.W. Lee, E.G. Radley and J.B. Simpson (HMSO, 1930) (lower) Craig Ellery; 31 (right and left) Alan McKirdy; 32 (upper) drawn by Jim Lewis (lower) Alan McKirdy; 33 John Gordon; 34 Brian Bell; 35 Stephen Finn/Alamy Stock Photo; 37 Pat & Angus Macdonald/Aerographica/SNH; 38 Pat & Angus Macdonald/Aerographica/SNH; 39 Lorne Gill/SNH; 40–1 (lower) Alan McKirdy; 41 (upper) Pat & Angus Macdonald/Aerographica/SNH; 42 (upper left) Aleksey Karpenko (upper right) Scott M. Ward (lower) Neil Burton; 43 (left) Sue Robinson (right) Lasse Ansaharju; 44 drawn by Jim Lewis; 45 (upper) Alexa Zari (left) Lorne Gill/SNH (right) Alan McKirdy; 46 (upper and lower) Alan McKirdy; 47 (left) Donna Carpenter (right) Alan McKirdy